Transformers: Basics, Maintenance, and Diagnostics

U.S. Department of the Interior
Bureau of Reclamation
Technical Service Center
Infrastructure Services Division
Hydroelectric Research and Technical Services Group
Denver, Colorado

April 2005

PREFACE

Transformers have been used at powerplants since the inception of alternating-current generation, a century ago. While operating principles of transformers remain the same, the challenges of maintaining and testing transformers have evolved along with transformer design and construction. Modern transformers are designed to closer tolerances than transformers in the past. Thus, effective, regular maintenance and testing is even more essential to continued operation when traditional "overdesign" cannot be relied on to overcome abnormal conditions. The utility engineer must be familiar with all aspects of maintenance and testing and make use of state-of-the-art tools and techniques for evaluating transformer condition. While on-line diagnostic systems and computerized testing methods are very helpful, they are not a substitute for sound engineering judgment and expertise.

This volume provides timely, practical advice to those seeking to better understand how transformers work, how they are best maintained, and how to test and evaluate their condition. It has been developed with the assistance of Bureau of Reclamation engineers responsible for operating and maintaining transformers at important powerplants in the Western States. Support and funding was provided through the Reclamation Power Resources Office in Denver and via the Manuals and Standards development program.

The authors gratefully acknowledge the assistance of all who contributed.

Hydroelectric Research and
Technical Services Group
Denver, Colorado
April 2005

Contents

	Page
1. Introduction	1
2. Introduction to Transformers	1
2.1 Principle of Operation	3
2.2 Transformer Action	4
2.3 Transformer Voltage and Current	8
2.4 The Magnetic Circuit	10
2.5 Core Losses	13
2.6 Copper Losses	13
2.7 Transformer Rating	14
2.8 Percent Impedance	14
2.9 Internal Forces	16
2.10 Autotransformers	17
2.11 Instrument Transformers	18
2.12 Potential Transformers	19
2.13 Current Transformers	20
2.14 Transformer Taps	22
2.15 Transformer Bushings	23
2.16 Transformer Polarity	24
2.17 Single-Phase Transformer Connections for Typical Service to Buildings	25
2.18 Parallel Operation of Single-Phase Transformers for Additional Capacity	26
2.19 Three-Phase Transformer Connections	28
2.20 Wye and Delta Connections	28
2.21 Three-Phase Connections Using Single-Phase Transformers	29
2.22 Paralleling Three-Phase Transformers	32
2.23 Methods of Cooling	35
2.24 Oil-Filled – Self-Cooled Transformers	36
2.25 Forced-Air and Forced-Oil-Cooled Transformers	36
2.26 Transformer Oil	37

Contents (continued)

Page

2. Introduction to Transformers (continued)
   2.27 Conservator System 38
   2.28 Oil-Filled, Inert-Gas System 39
   2.29 Indoor Transformers 41

3. Routine Maintenance 42
   3.1 Introduction to Reclamation Transformers 42
   3.2 Transformer Cooling Methods Introduction 44
   3.3 Dry-Type Transformers 45
      3.3.1 Potential Problems and Remedial Actions for Dry-Type Transformer Cooling Systems 47
   3.4 Liquid-Immersed Transformers 48
      3.4.1. Liquid-Immersed, Air-Cooled 48
      3.4.2 Liquid-Immersed, Air-Cooled/Forced Liquid-Cooled 50
      3.4.3 Liquid-Immersed, Water-Cooled 51
      3.4.4 Liquid-Immersed, Forced Liquid-Cooled 52
      3.4.5 Potential Problems and Remedial Actions for Liquid-Filled Transformer Cooling Systems 52
         3.4.5.1 Leaks 52
         3.4.5.2 Cleaning Radiators 53
         3.4.5.3 Plugged Radiators 53
         3.4.5.4 Sludge Formation 53
         3.4.5.5 Valve Problems 54
         3.4.5.6 Mineral Deposits 54
         3.4.5.7 Low Oil Level 54
      3.4.6 Cooling System Inspections 54

4. Oil-Filled Transformer Inspections 56
   4.1 Transformer Tank 57
   4.2 Top Oil Thermometers 57

Contents (continued)

Page

4. Oil-Filled Transformer Inspections (continued)
   4.3 Winding Temperature Thermometers........................ 58
      4.3.1 Temperature Indicators Online 60
      4.3.2 Temperature Indicators Offline...................... 60
   4.4 Oil Level Indicators 61
   4.5 Pressure Relief Devices 62
      4.5.1 Newer Pressure Relief Devices...................... 63
      4.5.2 Older Pressure Relief Devices 66
   4.6 Sudden Pressure Relay... 67
      4.6.1 Testing Suggestion.. 69
   4.7 Buchholz Relay (Found Only on Transformers with Conservators).. 69
   4.8 Transformer Bushings: Testing and Maintenance of High-Voltage Bushings................ 72
   4.9 Oil Preservation Sealing Systems 76
      4.9.1 Sealing Systems Types 77
         4.9.1.1 Free Breathing............................... 77
         4.9.1.2 Sealed or Pressurized Breathing 77
         4.9.1.3 Pressurized Inert Gas Sealed System.. 78
      4.9.2 Gas Pressure Control Components 81
         4.9.2.1 High-Pressure Gauge 81
         4.9.2.2 High-Pressure Regulator................. 81
         4.9.2.3 Low-Pressure Regulator................. 82
         4.9.2.4 Bypass Valve Assembly 82
         4.9.2.5 Oil Sump 82
         4.9.2.6 Shutoff Valves 83
         4.9.2.7 Sampling and Purge Valve............. 83
         4.9.2.8 Free Breathing Conservator 83
         4.9.2.9 Conservator with Bladder or Diaphragm Design 84
   4.10 Auxiliary Tank Sealing System 90

Contents (continued)

Page

5. Gaskets 91
   5.1 Sealing (Mating) Surface Preparation 92
   5.2 Cork-Nitrile 93
   5.3 Cork-Neoprene 94
   5.4 Nitrile "NBR" 95
      5.4.1 Viton 95
   5.5 Gasket Sizing for Standard Groove Depths 97
   5.6 Rectangular Nitrile Gaskets 99
   5.7 Bolting Sequences to Avoid Sealing Problems 103

6. Transformer Oils 105
   6.1 Transformer Oil Functions 105
      6.1.1 Dissolved Gas Analysis 105
      6.1.2 Key Gas Method 109
         6.1.2.1 Four-Condition DGA Guide (IEEE C57-104) 109
      6.1.3 Sampling Intervals and Recommended Actions 113
      6.1.4 Atmospheric Gases 117
      6.1.5 Dissolved Gas Software 117
      6.1.6 Temperature 122
      6.1.7 Gas Mixing 122
      6.1.8 Gas Solubility 123
      6.1.9 Diagnosing a Transformer Problem Using Dissolved Gas Analysis and the Duval Triangle 125
         6.1.9.1 Origin of the Duval Triangle 125
         6.1.9.2 How to Use the Duval Triangle 125
         6.1.9.3 Expertise Needed 131
         6.1.9.4 Rogers Ratio Method of DGA 131
      6.1.10 Carbon Dioxide/Carbon Monoxide Ratio 138
      6.1.11 Moisture Problems 141

Contents (continued)

Page

6. Transformer Oils (continued)
   6.1 Transformer Oil Functions (continued)
      6.1.11 Moisture Problems (continued)
         6.1.11.1 Dissolved Moisture in Transformer Oil 145
         6.1.11.2 Moisture in Transformer Insulation 146

7. Transformer Oil Tests that Should Be Completed Annually with the Dissolved Gas Analysis 150
   7.1 Dielectric Strength 150
      7.1.1 Interfacial Tension 151
   7.2 Acid Number 152
   7.3 Test for Oxygen Inhibitor 153
   7.4 Power Factor 154
   7.5 Oxygen 155
   7.6 Furans 155
   7.7 Oil Treatment Specification 159
      7.7.1 Taking Oil Samples for DGA 160
         7.7.1.1 DGA Oil Sample Container 162
         7.7.1.2 Taking the Sample 163

8. Silicone Oil-Filled Transformers 167
   8.1 Background 167
   8.2 Carbon Monoxide in Silicone Transformers 170
   8.3 Comparison of Silicone Oil and Mineral Oil Transformers 170
   8.4 Gas Limits 171
   8.5 Physical Test Limits 175

9. Transformer Testing 176
   9.1 DC Winding Resistance Measurement 176
   9.2 Core Insulation Resistance and Inadvertent Core Ground Test (Megger®) 178

Contents (continued)

Page

9. Transformer Testing (continued)
9.3 Doble Tests on Insulation 180
9.3.1 Insulation Power Factor Test 180
9.3.2 Capacitance Test 180
9.3.3 Excitation Current Test 181
9.3.4 Bushing Tests 182
9.3.5 Percent Impedance/Leakage Reactance Test 182
9.3.6 Sweep Frequency Response Analysis Tests 183
9.4 Visual Inspection 186
9.4.1 Background 186
9.4.2 Oil Leaks 187
9.4.3 Oil Pumps... 187
9.4.4 Fans and Radiators 188
9.4.5 Age 188
9.4.6 Infrared Temperature Analysis 189
9.4.7 IR for Transformer Tanks 189
9.4.8 IR for Surge Arresters 190
9.4.9 IR for Bushings 190
9.4.10 IR for Radiators and Cooling Systems.......... 191
9.4.11 Corona Scope Scan 193
9.5 Ultrasonic and Sonic Fault Detection 193
9.5.1 Background 193
9.5.2 Process 194
9.6 Vibration Analysis 194
9.6.1 Background 194
9.6.2 Process 195
9.7 Turns Ratio Test.. 195
9.7.1 Background 195
9.7.2 Process 195
9.8 Estimate of Paper Deterioration (Online) 196
9.8.1 CO_2 and CO Accumulated Total Gas Values 196

Contents (continued)

Page

9. Transformer Testing (continued)
    - 9.8 Estimate of Paper Deterioration (Online) (continued)
        - 9.8.2 CO_2/CO Ratio 196
    - 9.9 Estimate of Paper Deterioration (Offline During Internal Inspection) 197
        - 9.9.1 Degree of Polymerization (DP) 197
            - 9.9.1.1 Background 197
            - 9.9.1.2 Process 198
        - 9.9.2 Internal Inspection 198
            - 9.9.2.1 Background 198
            - 9.9.2.2 Transformer Borescope 200
    - 9.10 Transformer Operating History 200
    - 9.11 Transformer Diagnostics/Condition Assessment Summary 201

Appendix: Hydroplant Risk Assessment – Transformer Condition Assessment 205

References 231

Acronyms and Abbreviations 235

Tables

Table No. *Page*

1 Operative Parallel Connections of Three-Phase Transformers 34
2 Inoperative Parallel Connections of Three-Phase Transformers 34
3 Transformer Gasket Application Summary 96
4 Vertical Groove Compression for Circular Nitrile Gasket 97

Tables (continued)

Table No.		*Page*
5	Vertical Groove Compression for Rectangular Nitrile Gaskets	103
6	Transformer DGA Condition Summary Table	107
7	345-kV Transformer Example	108
8	Dissolved Key Gas Concentration Limits	110
9	Actions Based on Dissolved Combustible Gas	112
10	TOA L1 Limits and Generation Rate Per Month Alarm Limits	118
11	Fault Types	120
12	Dissolved Gas Solubility in Transformer Oil Accurate Only at STP, 0 °C (32 °F) and 14.7 psi (29.93 Inches of Mercury)	124
13	L1 Limits and Generation Rate Per Month Limits	126
14	Dissolved Gas Analysis Detection Limits	132
15	Rogers Ratios for Key Gases	135
16	Typical Faults in Power Transformers	140
17	Comparison of Water Distribution in Oil and Paper	142
18	Furans, DP, Percent of Life Used of Paper Insulation	156
19	Doble Limits for Inservice Oils	158
20	Additional Guidelines for Inservice Oils	159
21	Comparison of Gas Limits	172
22	Suggested Levels of Concern (Limits)	173
23	Doble and IEEE Physical Test Limits for Service-Aged Silicone Fluid	175
24	Paper Status Conditions Using CO_2 and CO	197
25	DP Values for Estimating Remaining Paper Life	198
26	Reclamation Transformer Condition Assessment Summary	201

Figures

Figure No. *Page*

1 Typical GSU Three-Phase Transformer 2
2 Transformer Construction 4
3 Transformer Action 6
4 Transformer 7
5 Step-Up and Step-Down Transformers 10
6 Magnetic Circuits 11
7 Three-Phase Core Form and Three-Phase Shell Form Transformer Units 12
8 Transformer Internal Forces 16
9 Autotransformers 17
10 Connections of Instrument Transformers 19
11 Potential Transformers 20
12 Current Transformer 21
13 Photograph of Current Transformers 22
14 Polarity Illustrated 25
15 Single-Phase Transformer 26
16 Single-Phase Paralleling 27
17 Three-Phase Connections 29
18 Delta-Delta Connections, Single-Phase Transformers for Three-Phase Operation 30
19 Wye-Wye Connections, Using Single-Phase Transformers for Three-Phase Operation 31
20 Delta-Wye and Wye-Delta Connections Using Single-Phase Transformers for Three-Phase Operation 33
21 Cooling 36
22 Forced-Air/Oil/Water-Cooled Transformers 37
23 Conservator with Bladder 38
24 Typical Transformer Nitrogen System 40
25 Transformer Diagnostics Flowchart 43
26 Typical Oil Flow 51
27 Oil Level Indicator 62
28 Conservator Oil Level 62

Figures (continued)

Figure No. *Page*

29 Pressure Relief Device 64
30 Photograph of a Pressure Relief Device 64
31 Sudden Pressure Relay, Section 68
32 Photograph of a Sudden Pressure Relay 69
33 Buchholz Relay, Section 70
34 Photograph of a Buchholz Relay 70
35 Pressurized Breathing Transformer 78
36 Pressurized Inert Gas Transformer 79
37 Gas Pressure Control Components 80
38 Free Breathing Conservator 84
39 Conservator with Bladder 85
40 Conservator Breather 87
41 Photograph of a Bladder Failure Relay 87
42 Bladder Failure Relay 88
43 Auxiliary Sealing System 91
44 Cross Section of Circular Gasket in Groove 98
45 Cross Section of Gasket Remains Constant Before Tightening and After 101
46 Bowing at Flanges 104
47 Bolt Tightening Sequences 104
48 Combustible Gas Generation Versus Temperature 115
49 The Duval Triangle 126
50 Duval Triangle Diagnostic Example of a Reclamation Transformer 128
51 Maximum Amount of Water Dissolved in Mineral Oil Versus Temperature 145
52 Transformer Oil Percent Saturation Curves 146
53 Water Distribution in Transformer Insulation 147
54 Myers Multiplier Versus Temperature 148
55 Water Content of Paper and Oil Nomogram 149
56 Interfacial Tension, Acid Number, Years in Service 152

Figures (continued)

Figure No. *Page*

57 Oil Sampling Piping 163
58 Sampling Syringe (Flushing) 164
59 Sampling Syringe (Filling) 165
60 Sample Syringe Bubble Removal 166
61 Relationship of Oxygen to Carbon Dioxide
and Carbon Monoxide as Transformer Ages 169
62 SFRA Test Traces of a New Transformer 184
63 SFRA Test Traces of a Defective New
Transformer.. 186
64 Normal Transformer IR Pattern 190
65 IR Image of Defective Arrester.................................. 191
66 IR Image Showing Blocked Radiators....................... 192
67 IR Image of Defective Bushing 192
68 Transformer Diagnostic Test Chart (Adapted
from IEEE 62-1995™) 204

1. Introduction

This document was created to provide guidance to Bureau of Reclamation (Reclamation) powerplant personnel in maintenance, diagnostics, and testing of transformers and associated equipment.

This document applies primarily to the maintenance and diagnostics of oil-filled power transformers (500 kilovoltamperes [kVA] and larger), owned and operated by Reclamation, although routine maintenance of other transformer types is addressed as well. Specific technical details are included in other documents and are referenced in this document.

Guidance and recommendations herein are based on industry standards and experience gained at Reclamation facilities. However, equipment and situations vary greatly, and sound engineering and management judgment must be exercised when applying these diagnostics. All available information must be considered (e.g., manufacturer's and transformer experts' recommendations, unusual operating conditions, personal experience with the equipment, etc.) in conjunction with this document.

2. Introduction to Transformers

Generator step-up (GSU) transformers represent the second largest capital investment in Reclamation power production—second only to generators. Reclamation has hundreds, perhaps thousands, of transformers, in addition to hundreds of large GSU transformers. Reclamation has transformers as small as a camera battery charger, about one-half the size of a coffee cup, to huge generator step-up transformers near the size of a small house. The total investment in transformers may well exceed generator investment. Transformers are extremely important to Reclamation, and it is necessary to understand their basic functions.

A transformer has no internal moving parts, and it transfers energy from one circuit to another by electromagnetic induction. External cooling may include heat exchangers, radiators, fans, and oil pumps. Radiators and fans are evident in figure 1. The large horizontal tank at the top is a conservator. Transformers are typically used because a change in voltage is needed. Power transformers are defined as transformers rated 500 kVA and larger. Larger transformers are oil-filled for insulation and cooling; a typical GSU transformer may contain several thousand gallons of oil. One must always be aware of the possibility of spills, leaks, fires, and environmental risks this oil poses.

Figure 1 – Typical GSU Three-Phase Transformer.

Transformers smaller than 500 kVA are generally called distribution transformers. Pole-top and small, pad-mounted transformers that serve residences and small businesses are typically distribution transformers. Generator step-up transformers, used in Reclamation powerplants, receive electrical energy at generator voltage and increase it to a higher voltage for transmission lines. Conversely, a step-down transformer receives energy at a higher voltage and delivers it at a lower voltage for distribution to various loads.

All electrical devices using coils (in this case, transformers) are constant wattage devices. This means voltage multiplied by current must remain constant; therefore, when voltage is "stepped-up," the current is "stepped-down" (and vice versa). Transformers transfer electrical energy between circuits completely insulated from each other. This makes it possible to use very high (stepped-up) voltages for transmission lines, resulting in a lower (stepped-down) current. Higher voltage and lower current reduce the required size and cost of transmission lines and reduce transmission losses as well. Transformers have made possible economic delivery of electric power over long distances.

Transformers do not require as much attention as most other equipment; however, the care and maintenance they do require is absolutely critical. Because of their reliability, maintenance is sometimes ignored, causing reduced service life and, at times, outright failure.

2.1 Principle of Operation

Transformer function is based on the principle that electrical energy is transferred efficiently by magnetic induction from one circuit to another. When one winding of a transformer is energized from an alternating current (AC) source, an alternating magnetic field is established in the transformer core. Alternating magnetic lines of force, called "flux," circulate through the core. With a second winding around the same core, a voltage is induced by the alternating flux lines. A circuit, connected to the terminals of the second winding, results in current flow.

Each phase of a transformer is composed of two separate coil windings wound on a common core. The low-voltage winding is placed nearest the core; the high-voltage winding is then placed around both the low-voltage winding and core. See figure 2 which shows internal construction of one phase. The core is typically made from very thin steel laminations, each coated with insulation. By insulating between individual laminations, losses are reduced. The steel core provides a low resistance path for magnetic flux. Both high- and low-voltage windings are insulated from the core and from each other, and leads

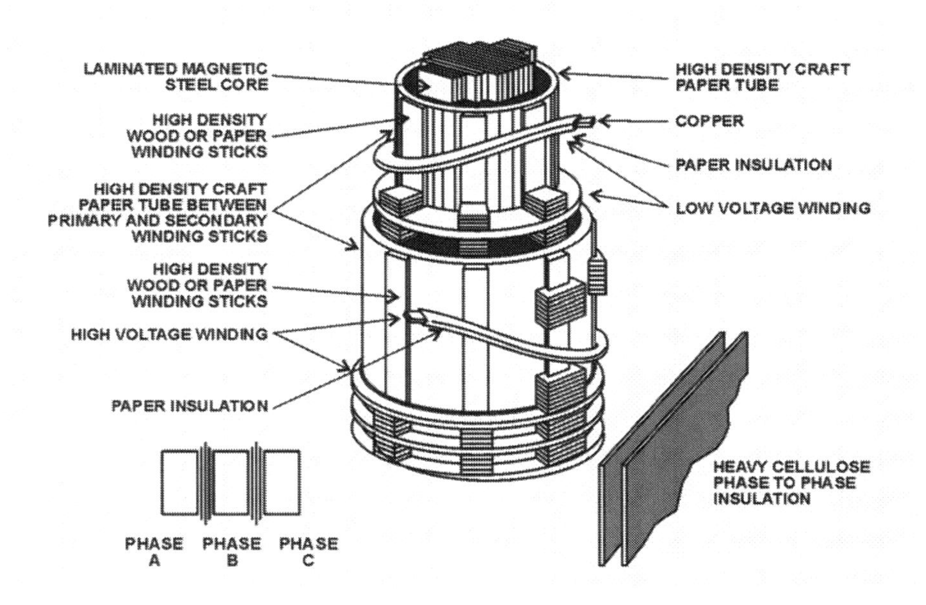

Figure 2 – Transformer Construction.

are brought out through insulating bushings. A three-phase transformer typically has a core with three legs and has both high-voltage and low-voltage windings around each leg. Special paper and wood are used for insulation and internal structural support.

2.2 Transformer Action

Transformer action depends upon magnetic lines of force (flux) mentioned above. At the instant a transformer primary is energized with AC, a flow of electrons (current) begins. During the instant of switch closing, buildup of current and magnetic field occurs. As current begins the positive portion of the sine wave, lines of magnetic force (flux) develop outward from the coil and continue to expand until the current is at its positive peak. The magnetic field is also at its positive peak. The current sine wave then begins to decrease, crosses zero, and goes negative until it reaches its negative peak. The magnetic flux switches direction and also reaches its peak in the opposite direction. With an AC power circuit, the current changes (alternates) continually 60 times per second, which is standard in the United States. Other countries may use other frequencies. In Europe, 50 cycles per second is common.

Strength of a magnetic field depends on the amount of current and number of turns in the winding. When current is reduced, the magnetic field shrinks. When the current is switched off, the magnetic field collapses.

When a coil is placed in an AC circuit, as shown in figure 3, current in the primary coil will be accompanied by a constantly rising and collapsing magnetic field. When another coil is placed within the alternating magnetic field of the first coil, the rising and collapsing flux will induce voltage in the second coil.

When an external circuit is connected to the second coil, the induced voltage in the coil will cause a current in the second coil. The coils are said to be magnetically coupled; they are, however, electrically isolated from each other.

Many transformers have separate coils, as shown in figure 3, and contain many turns of wire and a magnetic core, which forms a path for and concentrates the magnetic flux. The winding receiving electrical energy from the source is called the primary winding. The winding which receives energy from the primary winding, via the magnetic field, is called the "secondary" winding.

Either the high- or low-voltage winding can be the primary or the secondary. With GSUs at Reclamation powerplants, the primary winding is the low-voltage side (generator voltage), and the high-voltage side is the secondary winding (transmission voltage). Where power is used (i.e., at residences or businesses), the primary winding is the high-voltage side, and the secondary winding is the low-voltage side.

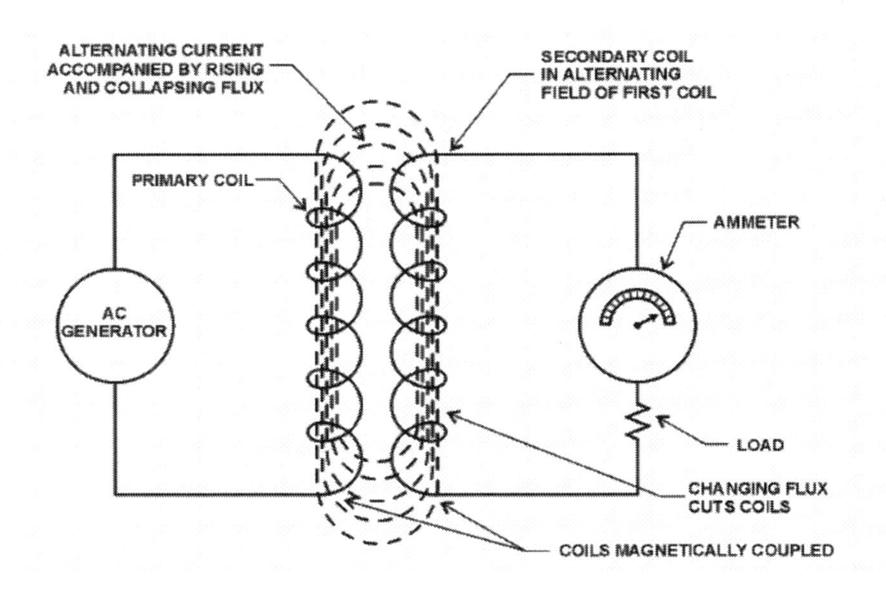

Figure 3 – Transformer Action.

The amount of voltage induced in each turn of the secondary winding will be the same as the voltage across each turn of the primary winding. The total amount of voltage induced will be equal to the sum of the voltages induced in each turn. Therefore, if the secondary winding has more turns than the primary, a greater voltage will be induced in the secondary; and the transformer is known as a step-up transformer. If the secondary winding has fewer turns than the primary, a lower voltage will be induced in the secondary; and the transformer is a step-down transformer. Note that the primary is always connected to the source of power, and the secondary is always connected to the load.

In actual practice, the amount of power available from the secondary will be slightly less than the amount supplied to the primary because of losses in the transformer itself.

When an AC generator is connected to the primary coil of a transformer (figure 4), electrons flow through the coil due to the

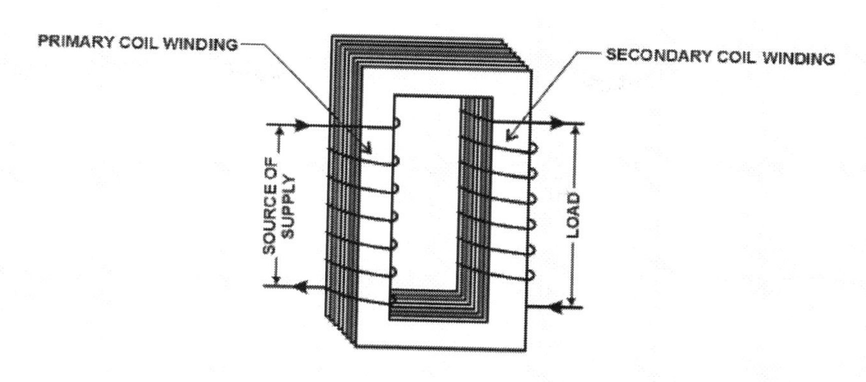

Figure 4 – Transformer.

generator voltage. Alternating current varies, and accompanying magnetic flux varies, cutting both transformer coils and inducing voltage in each coil circuit.

The voltage induced in the primary circuit opposes the applied voltage and is known as back voltage or back electro-motive-force (back EMF). When the secondary circuit is open, back EMF, along with the primary circuit resistance, acts to limit the primary current. Primary current must be sufficient to maintain enough magnetic field to produce the required back EMF.

When the secondary circuit is closed and a load is applied, current appears in the secondary due to induced voltage, resulting from flux created by the primary current. This secondary current sets up a second magnetic field in the transformer in the opposite direction of the primary field. Thus, the two fields oppose each other and result in a combined magnetic field of less strength than the single field produced by the primary with the secondary open. This reduces the back voltage (back EMF) of the primary and causes the primary current to increase. The primary current increases until it re-establishes the total magnetic field at its original strength.

In transformers, a balanced condition must always exist between the primary and secondary magnetic fields. Volts times amperes (amps) must also be balanced (be the same) on both primary and secondary. The required primary voltage and current must be supplied to maintain the transformer losses and secondary load.

2.3 Transformer Voltage and Current

If the small amount of transformer loss is ignored, the back-voltage (back EMF) of the primary must equal the applied voltage. The magnetic field, which induces the back-voltage in the primary, also cuts the secondary coil. If the secondary coil has the same number of turns as the primary, the voltage induced in the secondary will equal the back-voltage induced in the primary (or the applied voltage). If the secondary coil has twice as many turns as the primary, it will be cut twice as many times by the flux, and twice the applied primary voltage will be induced in the secondary. The total induced voltage in each winding is proportional to the number of turns in that winding. If E_1 is the primary voltage and I_1 the primary current, E_2 the secondary voltage and I_2 the secondary current, N_1 the primary turns and N_2 the secondary turns, then:

$$\frac{E_1}{E_2} = \frac{N_1}{N_2} = \frac{I_2}{I_1}$$

Note that the current is inversely proportional to both voltage and number of turns. This means (as discussed earlier) that if voltage is stepped up, the current must be stepped down and vice versa. The number of turns remains constant unless there is a tap changer (discussed later).

The power output or input of a transformer equals volts times amperes (E x I). If the small amount of transformer loss is disregarded, input equals output or:

$$E_1 \times I_1 = E_2 \times I_2$$

If the primary voltage of a transformer is 110 volts (V), the primary winding has 100 turns, and the secondary winding has 400 turns, what will the secondary voltage be?

$$\frac{E_1}{E_2} = \frac{N_1}{N_2} \qquad \frac{110}{E_2} = \frac{100}{400}$$

$$100\ E_2 = 44{,}000 \qquad E_2 = 440 \text{ volts}$$

If the primary current is 20 amps, what will the secondary current be?

$$E_2 \times I_2 = E_1 \times I_1$$

$$440 \times I_2 = 110 \times 20 = 2{,}200$$

$$I_2 = 5 \text{ amps}$$

Since there is a ratio of 1 to 4 between the turns in the primary and secondary circuits, there must be a ratio of 1 to 4 between the primary and secondary voltage and a ratio of 4 to 1 between the primary and secondary current. As voltage is stepped up, the current is stepped down, keeping volts multiplied by amps constant. This is referred to as "volt amps."

As mentioned earlier and further illustrated in figure 5, when the number of turns or voltage on the secondary of a transformer is greater than that of the primary, it is known as a step-up transformer. When the number of turns or voltage on the secondary is less than that of the primary, it is known as a step-down transformer. A power transformer used to tie two systems together may feed current either way between systems, or act as a step-up or step-down transformer, depending on where power is being generated and where it is consumed. As mentioned above, either winding could be the primary or secondary. To eliminate this confusion, in power generation, windings of transformers are often referred to as high-side and low-side windings, depending on the relative values of the voltages.

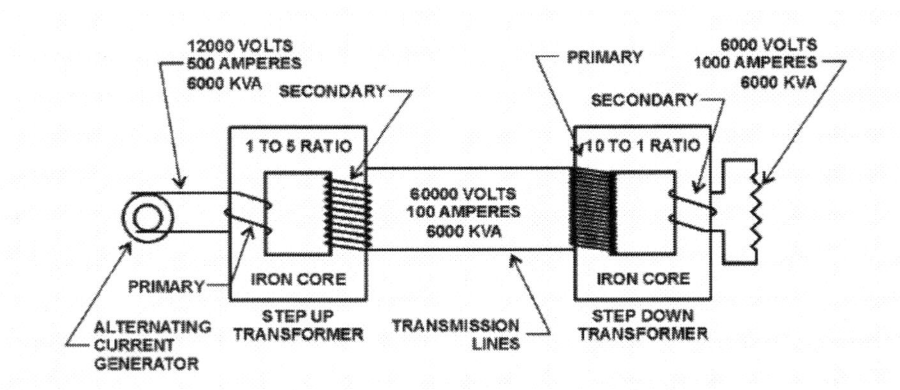

Figure 5 – Step-Up and Step-Down Transformers.

Note that kVA (amps times volts) remains constant throughout the above circuit on both sides of each transformer, which is why they are called constant wattage devices.

Efficiencies of well-designed power transformers are very high, averaging over 98 percent (%). The only losses are due to core losses, maintaining the alternating magnetic field, resistance losses in the coils, and power used for cooling. The main reason for high efficiencies of power transformers, compared to other equipment, is the absence of moving parts. Transformers are called static AC machines.

2.4 The Magnetic Circuit

A magnetic circuit or core of a transformer is designed to provide a path for the magnetic field, which is necessary for induction of voltages between windings. A path of low reluctance (i.e., resistance to magnetic lines of force), consisting of thin silicon, sheet steel laminations, is used for this purpose. In addition to providing a low reluctance path for the magnetic field, the core is designed to prevent circulating electric currents within the steel itself. Circulating currents, called eddy currents, cause heating and energy loss. They are due to voltages induced in the steel of the core, which is constantly subject to alternating magnetic fields. Steel itself is a conductor, and changing lines of magnetic flux also induce a voltage and current in this conductor. By using very thin sheets of steel with insulating material between sheets, eddy currents (losses) are greatly reduced.

The two common arrangements of the magnetic path and the windings are shown in figure 6 and 7. In the core-type (core form) transformer, the windings surround the core.

A section of both primary and secondary windings are wound on each leg of the core, the low voltage winding is wound next to the core, and the high voltage winding is wound over the low voltage.

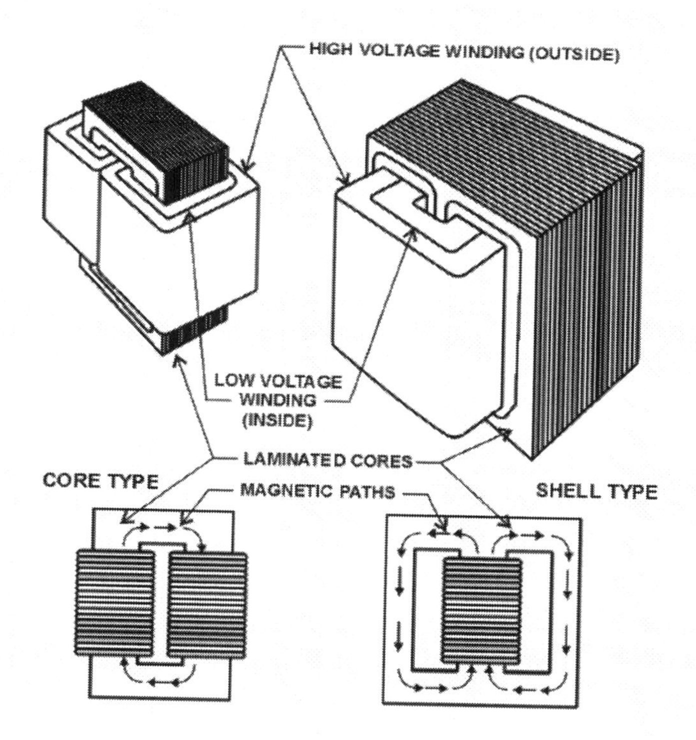

Figure 6 – Magnetic Circuits.

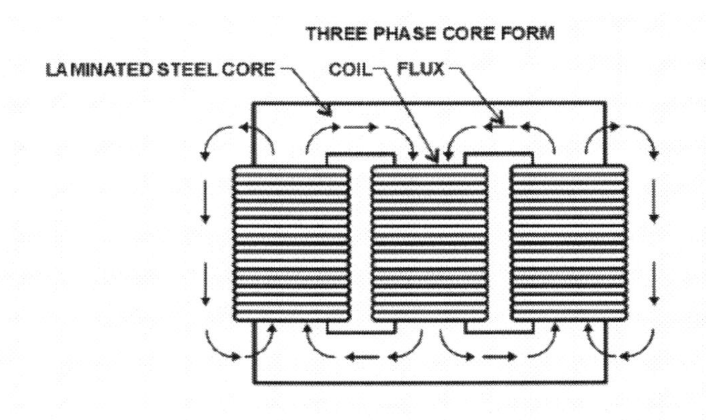

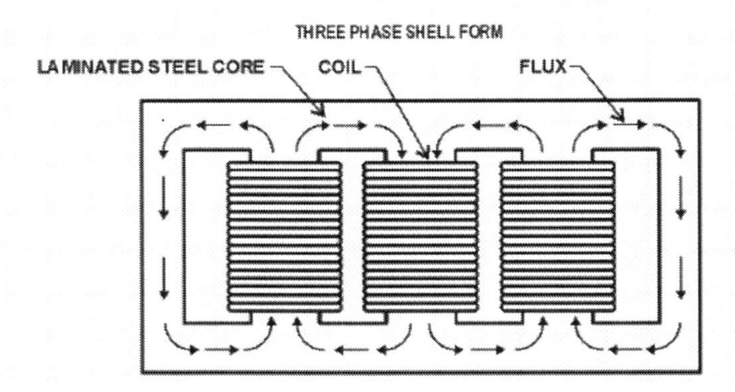

Figure 7 – Three-Phase Core Form and Three-Phase Shell Form Transformer Units.

In a shell-type (shell form) transformer, the steel magnetic circuit (core) forms a shell surrounding the windings. In a core form, the windings are on the outside; in a shell form, the windings are on the inside. In power transformers, the electrical windings are arranged so that practically all of the magnetic lines of force go through both the primary and secondary windings. A small percentage of the magnetic lines of force goes outside the core, and this is called leakage flux. Larger transformers, such as Reclamation GSU transformers, are almost always shell type.

Note that, in the shell form transformers, (see figure 7) the magnetic flux, external to the coils on both left and right extremes, has complete magnetic paths for stray and zero sequence flux to return to the coils. In the core form, it can easily be seen from the figure that, on both left and right extremes, there are no return paths. This means that the flux must use external tank walls and the insulating medium for return paths. This increases core losses and decreases overall efficiency and shows why most large transformers are built as shell form units.

2.5 Core Losses

Since magnetic lines of force in a transformer are constantly changing in value and direction, heat is developed because of the hysteresis of the magnetic material (friction of the molecules). This heat must be removed; therefore, it represents an energy loss of the transformer. High temperatures in a transformer will drastically shorten the life of insulating materials used in the windings and structures. For every 8 degrees Celsius (°C) temperature rise, life of the transformer is cut by one-half; therefore, maintenance of cooling systems is critical.

Losses of energy, which appears as heat due both to hysteresis and to eddy currents in the magnetic path, is known as core losses. Since these losses are due to alternating magnetic fields, they occur in a transformer whenever the primary is energized, even though no load is on the secondary winding.

2.6 Copper Losses

There is some loss of energy in a transformer due to resistance of the primary winding to the magnetizing current, even when no load is connected to the transformer. This loss appears as heat generated in the winding and must also be removed by the cooling system. When a load is connected to a transformer and electrical currents exist in both primary and secondary windings, further losses of electrical energy occur. These losses, due to resistance of the windings, are called copper losses (or the I^2R losses).

2.7 Transformer Rating

Capacity (or rating) of a transformer is limited by the temperature that the insulation can tolerate. Ratings can be increased by reducing core and copper losses, by increasing the rate of heat dissipation (better cooling), or by improving transformer insulation so it will withstand higher temperatures. A physically larger transformer can dissipate more heat, due to the increased area and increased volume of oil. A transformer is only as strong as its weakest link, and the weakest link is the paper insulation, which begins to degrade around 100 °C. This means that a transformer must be operated with the "hottest spot" cooler than this degradation temperature, or service life is greatly reduced. Reclamation typically orders transformers larger than required, which aids in heat removal and increases transformer life.

Ratings of transformers are obtained by simply multiplying the current times the voltage. Small transformers are rated in "VA," volts times amperes. As size increases, 1 kilovoltampere (kVA) means 1,000 voltamperes, 1 megavoltampere (MVA) means 1 million voltamperes. Large GSUs may be rated in hundreds of MVAs. A GSU transformer can cost well over a million dollars and take 18 months to 2 years or longer to obtain. Each one is designed for a specific application. If one fails, this may mean a unit or whole plant could be down for as long 2 years, costing multiple millions of dollars in lost generation, in addition to the replacement cost of the transformer itself. This is one reason that proper maintenance is critical.

2.8 Percent Impedance

The impedance of a transformer is the total opposition offered an alternating current. This may be calculated for each winding. However, a rather simple test provides a practical method of measuring the equivalent impedance of a transformer without separating the impedance of the windings. When referring to impedance of a transformer, it is the equivalent impedance that is meant. In order to determine equivalent impedance, one winding of the transformer is short circuited, and just enough voltage is applied to the other winding to create full load current to flow in the short circuited winding. This voltage is known as the impedance voltage. Either winding may be short-circuited for this test, but it is usually more convenient to short circuit the low-voltage winding. The transformer impedance value is given on the nameplate in percent. This means that the voltage drop due to the impedance is expressed as a percent of rated voltage. For example, if a 2,400/240-volt transformer has a measured impedance voltage of 72 volts on the high-voltage windings, its impedance (Z), expressed as a percent, is:

$$\text{percent } Z = \frac{72}{2{,}400} \times 100 = 3 \text{ percent}$$

This means there would be a 72-volt drop in the high-voltage winding at full load due to losses in the windings and core. Only 1 or 2% of the losses are due to the core; about 98% are due to the winding impedance. If the transformer were not operating at full load, the voltage drop would be less. If an actual impedance value in ohms is needed for the high-voltage side:

$$Z = \frac{V}{I}$$

where V is the voltage drop or, in this case, 72 volts; and I is the full load current in the primary winding. If the full load current is 10 amps:

$$Z = \frac{72\,V}{10\,a} = 7.2\ ohms$$

Of course, one must remember that impedance is made up of both resistive and reactive components.

2.9 Internal Forces

During normal operation, internal structures and windings are subjected to mechanical forces due to the magnetic forces. These forces are illustrated in figure 8. By designing the internal structure very strong to withstand these forces over a long period of time, service life can be extended. However, in a large transformer during a "through fault" (fault current passing through a transformer), forces can reach millions of pounds, pulling the coils up and down and pulling them apart 60 times per second. Notice in figure 8 that the internal low-voltage coil is being pulled downward, while the high-voltage winding is pulled up, in the opposite direction. At the same time, the right-hand part of the figure shows that the high- and low-voltage coils are being forced apart. Keep in mind that these forces are reversing 60 times each second. It is obvious why internal structures of transformers must be built incredibly strong.

Many times, if fault currents are high, these forces can rip a transformer apart and cause electrical faults inside the transformer itself. This normally results in arcing inside the transformer that can result in explosive failure of the tank, throwing flaming oil over a wide

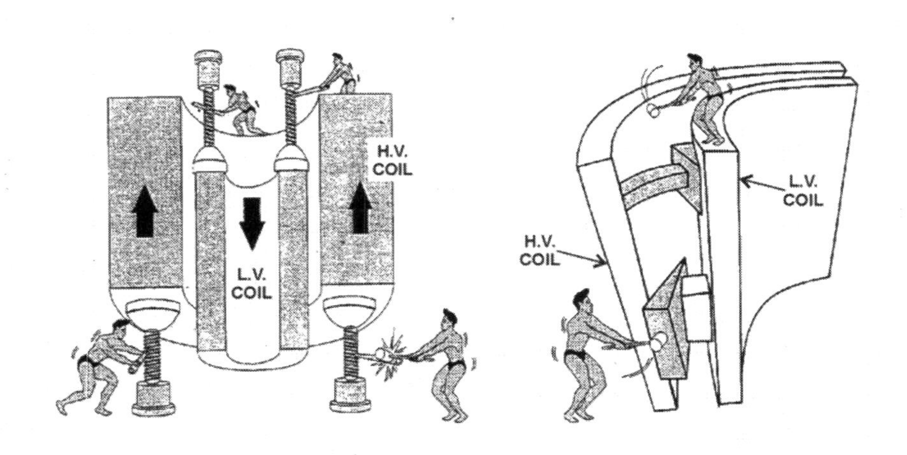

Figure 8 – Transformer Internal Forces.

area. There are protective relaying systems to protect against this possibility, although explosive failures do occur occasionally.

2.10 Autotransformers

It is possible to obtain transformer action by means of a single coil, provided that there is a "tap connection" somewhere along the winding. Transformers having only one winding are called autotransformers, shown schematically in figure 9.

An autotransformer has the usual magnetic core but only one winding, which is common to both the primary and secondary circuits.

The primary is always the portion of the winding connected to the AC power source. This transformer may be used to step voltage up or down. If the primary is the total winding and is connected to a supply, and the secondary circuit is connected across only a portion of the winding (as shown), the secondary voltage is "stepped-down."

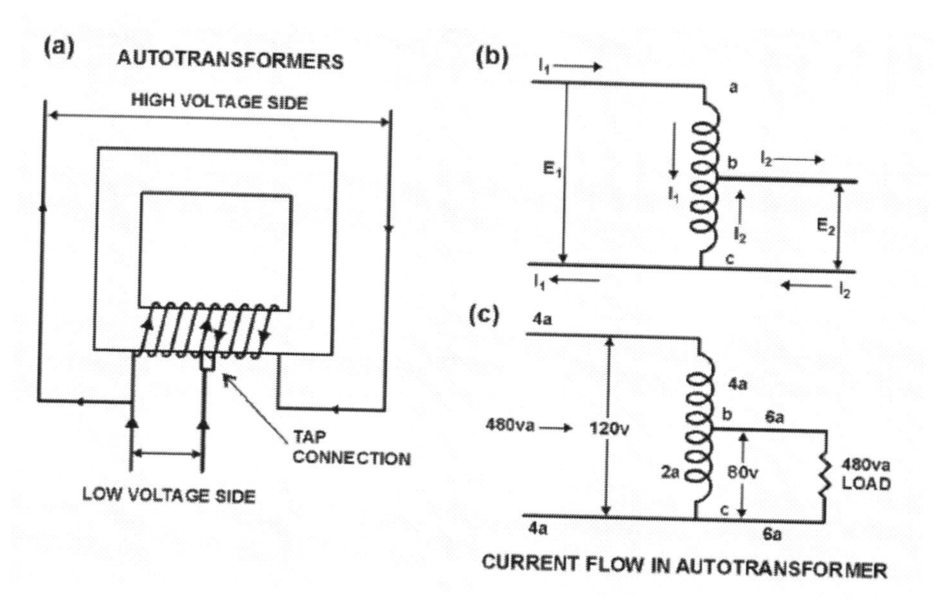

Figure 9 – Autotransformers.

If only a portion of the winding is the primary and is connected to the supply voltage and the secondary includes all the winding, then the voltage will be "stepped-up" in proportion to the ratio of the total turns to the number of connected turns in the primary winding.

When primary current I_1 is in the direction of the arrow, secondary current, I_2, is in the opposite direction, as in figure 9b. Therefore, in the portion of the winding between points b and c, current is the difference of I_1 and I_2. If the requirement is to step the voltage up (or down) only a small amount, then the transformer ratio is small—E_1 and E_2 are nearly equal. Currents I_1 and I_2 are also nearly equal. The portion of the winding between b and c, which carries the difference of the currents, can be made of a much smaller conductor, since the current is much lower.

Under these circumstances, the autotransformer is much cheaper than the two-coil transformer of the same rating. However, the disadvantage of the autotransformer is that the primary and secondary circuits are electrically connected and, therefore, could not safely be used for stepping down from high voltage to a voltage suitable for plant loads. The autotransformer, however, is extensively used for reducing line voltage for step increases in starting larger induction motors. There are generally four or five taps that are changed by timers so that more of the winding is added in each step until the full voltage is applied across the motor. This avoids the large inrush current required when starting motors at full line voltage. This transformer is also extensively used for "buck-boost" when the voltage needs to be stepped up or down only a small percentage. One very common example is boosting 208 V up from one phase of a 120/208-V three-phase system, to 220 V for single-phase loads.

2.11 Instrument Transformers

Instrument transformers (figure 10) are used for measuring and control purposes. They provide currents and voltages proportional to the primary, but there is less danger to instruments and personnel.

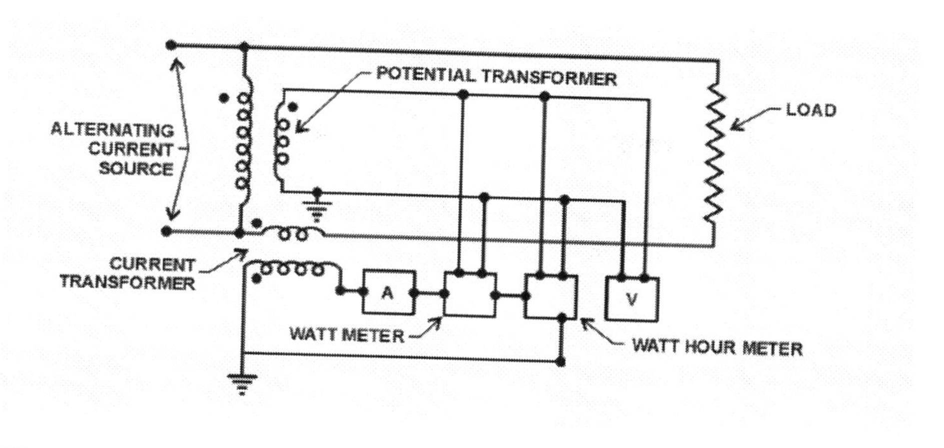

Figure 10 – Connections of Instrument Transformers.

Those transformers used to step voltage down are known as potential transformers (PTs) and those used to step current down are known as current transformers (CTs).

The function of a PT is to accurately measure voltage on the primary, while a CT is used to measure current on the primary.

2.12 Potential Transformers

Potential transformers (figure 11) are used with voltmeters, wattmeters, watt-hour meters, power-factor meters, frequency meters, synchroscopes and synchronizing apparatus, protective and regulating relays, and undervoltage and overvoltage trip coils of circuit breakers. One potential transformer can be used for a number of instruments if the total current required by the instruments connected to the secondary winding does not exceed the transformer rating.

Potential transformers are usually rated 50 to 200 volt-amperes at 120 secondary volts. The secondary terminals should never be short circuited because a heavy current will result, which can damage the windings.

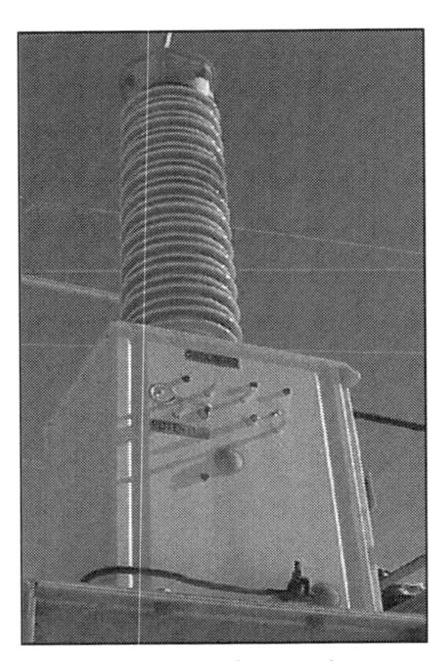

Potential Transformer in switchyard.

Inside Potential Transformer with Fuses.

Figure 11 – Potential Transformers.

2.13 Current Transformers

The primary of a current transformer typically has only one turn. This is not really a turn or wrap around the core but just a conductor or bus going through the "window." The primary never has more than a very few turns, while the secondary may have a great many turns, depending upon how much the current must be stepped down. In most cases, the primary of a current transformer is a single wire or bus bar, and the secondary is wound on a laminated magnetic core, placed around the conductor in which the current needs to be measured, as illustrated in figure 12.

If primary current exists and the secondary circuit of a CT is closed, the winding builds and maintains a counter or back EMF to the primary magnetizing force. Should the secondary be opened with current in the primary, the counter EMF is removed; and the primary magnetizing force builds up an extremely high

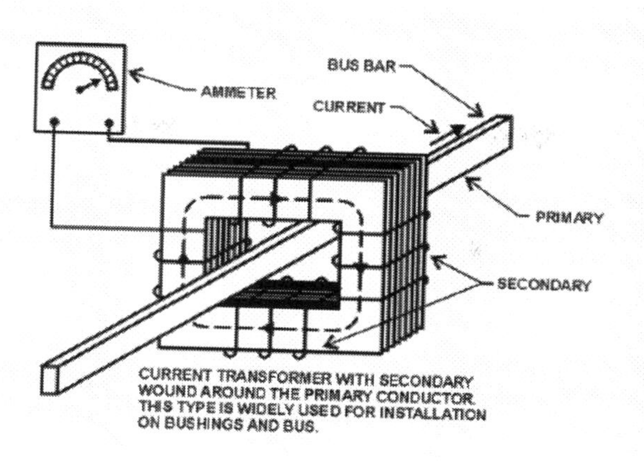

Figure 12 – Current Transformer.

voltage in the secondary, which is dangerous to personnel and can destroy the current transformer.

CAUTION:

For this reason, the secondary of a current transformer should always be shorted before removing a relay from its case or removing any other device that the CT operates. This protects the CT from overvoltage.

Current transformers are used with ammeters, wattmeters, power-factor meters, watt-hour meters, compensators, protective and regulating relays, and trip coils of circuit breakers. One CT can be used to operate several instruments, provided the combined loads of the instruments do not exceed that for which the CT is rated. Secondary windings are usually rated at 5 amperes. A variety of current transformers are shown in figure 13. Many times, CTs have several taps on the secondary winding to adjust the range of current possible to measure on the primary.

Current Transformer in Switchyard.

Lab Current Transformer.

Current Transformer in Switchyard.

Figure 13 – Photograph of Current Transformers.

2.14 Transformer Taps

Most power transformers have taps on either primary or secondary windings to vary the number of turns and, thus, the output voltage. The percentage of voltage change, above or below normal, between different tap positions varies in different transformers. In oil-cooled transformers, tap leads are brought to a tap changer, located beneath the oil inside the tank, or brought to an oil-filled tap changer, externally located. Taps on dry-type transformers are brought to insulated terminal boards located inside the metal housing, accessible by removing a panel.

Some transformers taps can be changed under load, while other transformers must be de-energized. When it is necessary to change taps frequently to meet changing conditions, taps that can be changed under load are used. This is accomplished by means of a motor that may be controlled either manually or automatically. Automatic operation is achieved by changing taps to maintain constant voltage as system conditions change. A common range of adjustment is plus or minus 10%. At Reclamation powerplants, de-energized tap changers (DETC) are used and can only be changed with the transformer off-line. A very few load tap changers (LTC) are used at Grand Coulee between the 500-kilovolt (kV) (volts x 1,000) and 220-kV switchyards.

A bypass device is sometimes used across tap changers to ensure power flow in case of contact failure. This prevents failure of the transformer in case excessive voltage appears across faulty contacts.

2.15 Transformer Bushings

The two most common types of bushings used on transformers as main lead entrances are solid porcelain bushings on smaller transformers and oil-filled condenser bushings on larger transformers.

Solid porcelain bushings consist of high-grade porcelain cylinders that conductors pass through. Outside surfaces have a series of skirts to increase the leakage path distance to the grounded metal case. High-voltage bushings are generally oil-filled condenser type. Condenser types have a central conductor wound with alternating layers of paper insulation and tin foil and filled with insulating oil. This results in a path from the conductor to the grounded tank, consisting of a series of condensers. The layers are designed to provide approximately equal voltage drops between each condenser layer.

Acceptance and routine maintenance tests most often used for checking the condition of bushings are Doble power factor tests. The power factor of a bushing in good condition will remain relatively stable throughout the service life. A good indication of insulation deterioration is a slowly rising power factor. The most common cause of failure is moisture entrance through the top bushing seal. This condition will be revealed before failure by routine Doble testing. If Doble testing is not performed regularly, explosive failure is the eventual result of a leaking bushing. This, many times, results in a catastrophic and expensive failure of the transformer as well.

2.16 Transformer Polarity

With power or distribution transformers, polarity is important only if the need arises to parallel transformers to gain additional capacity or to hook up three single-phase transformers to make a three-phase bank. The way the connections are made affects angular displacement, phase rotation, and direction of rotation of connected motors. Polarity is also important when hooking up current transformers for relay protection and metering. Transformer polarity depends on which direction coils are wound around the core (clockwise or counterclockwise) and how the leads are brought out. Transformers are sometimes marked at their terminals with polarity marks. Often, polarity marks are shown as white paint dots (for plus) or plus-minus marks on the transformer and symbols on the nameplate. These marks show the connections where the input and output voltages (and currents) have the same instantaneous polarity.

More often, transformer polarity is shown simply by the American National Standards Institute (ANSI) designations of the winding leads as H_1, H_2 and X_1, X_2. By ANSI standards, if you face the low-voltage side of a single-phase transformer (the side marked X_1, X_2), the H_1 connection will always be on your far left. See the single-phase diagrams in figure 14. If the terminal marked X_1 is also on your left, it is subtractive polarity. If the X_1 terminal is on your right, it is additive polarity. Additive polarity is common for small distribution transformers. Large transformers, such as GSUs at Reclamation powerplants, are generally subtractive polarity.

It is also helpful to think of polarity marks in terms of current direction. At any instant when the current direction is into a polarity marked terminal of the primary winding, the current direction is out of the terminal with the same polarity mark in the secondary winding. It is the same as if there were a continuous circuit across the two windings.

Polarity is a convenient way of stating how leads are brought out. If you want to test for polarity, connect the transformer as shown

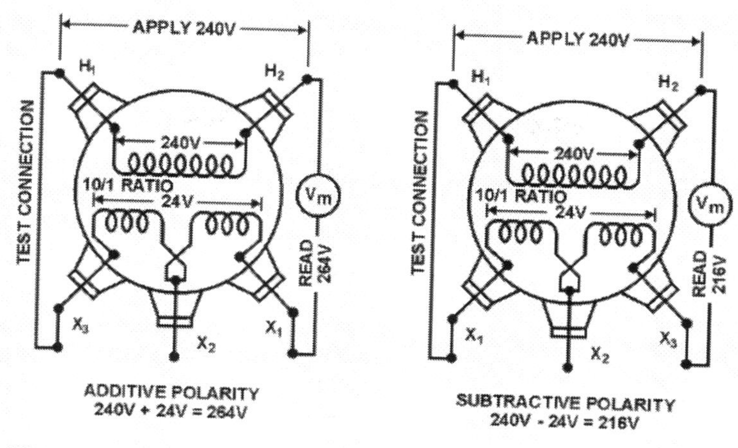

Figure 14 – Polarity Illustrated.

in figure 14. A transformer is said to have additive polarity if, when adjacent high- and low-voltage terminals are connected and a voltmeter placed across the other high- and low-voltage terminals, the voltmeter reads the sum (additive) of the high- and low-voltage windings. It is subtractive polarity if the voltmeter reads the difference (subtractive) between the voltages of the two windings. If this test is conducted, use the lowest AC voltage available to reduce potential hazards. An adjustable ac voltage source, such as a variac, is recommended to keep the test voltage low.

2.17 Single-Phase Transformer Connections for Typical Service to Buildings

Figure 15 shows a typical arrangement of bringing leads out of a single-phase distribution transformer. To provide flexibility for connection, the secondary winding is arranged in two sections.

Each section has the same number of turns and, consequently, the same voltage. Two primary leads (H_1, H_2) are brought out from the

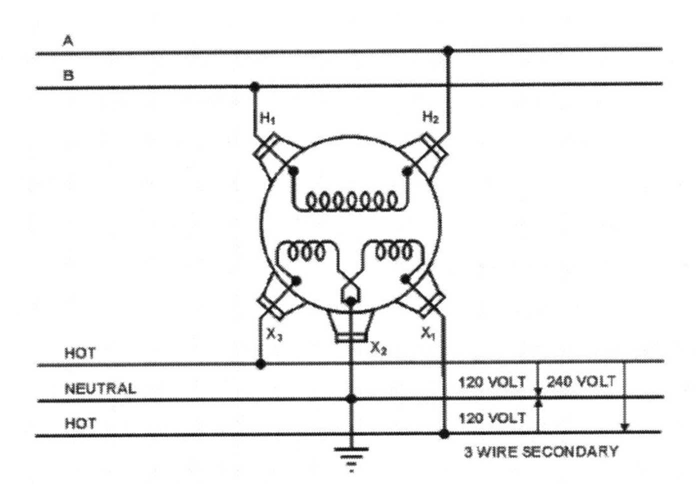

Figure 15 – Single-Phase Transformer.

top through porcelain bushings. Three secondary leads (X_1, X_2, X_3) are brought out through insulating bushings on the side of the tank, one lead from the center tap (neutral) (X_2) and one from each end of the secondary coil (X_1 and X_3). Connections, as shown, are typical of services to homes and small businesses. This connection provides a three-wire service that permits adequate capacity at minimum cost. The neutral wire (X_2) (center tap) is grounded. A 120-volt circuit is between the neutral and each of the other leads, and a 240-volt circuit is between the two ungrounded leads.

2.18 Parallel Operation of Single-Phase Transformers for Additional Capacity

In perfect parallel operation of two or more transformers, current in each transformer would be directly proportional to the transformer capacity, and the arithmetic sum would equal one-half the total current. In practice, this is seldom achieved because of small variations in transformers. However, there are conditions for operating transformers in parallel. They are:

1. Any combination of positive and negative polarity transformers can be used. However, in all cases, numerical notations **must be followed** on both primary and secondary connections. That is H_1 connected to H_1, H_2 connected to H_2, and X_1 connected to X_1, X_2 connected to X_2, X_3 connected to X_3. Note that each subscript number on a transformer must be connected to the same subscript number on the other transformer as shown in figure 16.

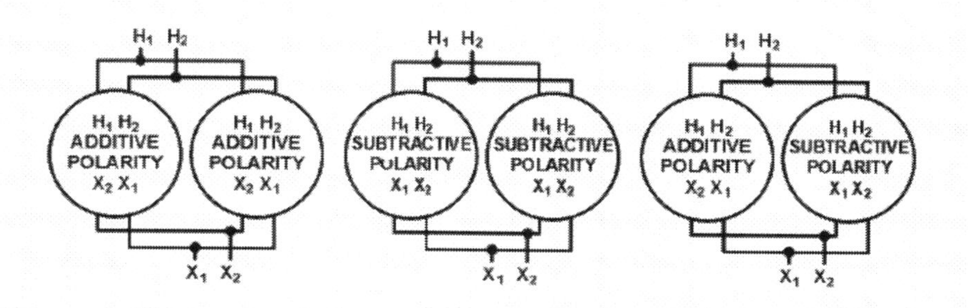

Figure 16 – Single-Phase Paralleling.

CAUTION:

With positive and negative polarity transformers, the location of X_1 and X_2 connections on the tanks will be reversed. Care must be exercised to ensure that terminals are connected, as stated above. See figure 16.

---

2. Tap settings must be identical.
3. Voltage ratings must be identical; this, of course, makes the turns ratios also identical.
4. The percent impedance of one transformer must be between 92½% and 107½% of the other. Otherwise, circulating currents between the two transformers would be excessive.

5. Frequencies must be identical. Standard frequency in the United States is 60 hertz and usually will not present a problem.

One will notice, from the above requirements, that paralleled transformers do not have to be the same size. However, to meet the percent impedance requirement, they must be nearly the same size. Most utilities will not parallel transformers if they are more than one standard kVA size rating different from each other; otherwise, circulating currents are excessive.

2.19 Three-Phase Transformer Connections

Three-phase power is attainable with one three-phase transformer, which is constructed with three single-phase units enclosed in the same tank or three separate single-phase transformers. The methods of connecting windings are the same, whether using the one three-phase transformer or three separate single-phase transformers.

2.20 Wye and Delta Connections

The two common methods of connecting three-phase generators, motors, and transformers are shown in figure 17. The method shown in at figure 17a is known as a delta connection, because the diagram bears a close resemblance to the Greek letter Δ, called delta.

The other method, figure 17b, is known as the star or wye connection. The wye differs from the delta connection in that it has two phases in series. The common point "O" of the three windings is called the neutral because equal voltages exist between this point and any of the three phases.

When windings are connected wye, the voltage between any two lines will be 1.732 times the phase voltage, and the line current will be the same as the phase current. When transformers are connected delta, the line current will be 1.732 times the phase current, and the voltage between any two will be the same as that of the phase voltage.

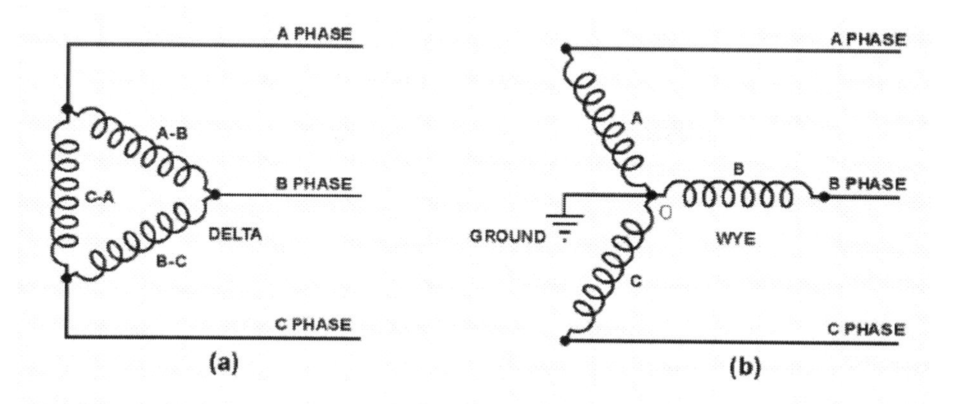

Figure 17 – Three-Phase Connections.

2.21 Three-Phase Connections Using Single-Phase Transformers

As mentioned above, single-phase transformers may be connected to obtain three-phase power. These are found at many Reclamation facilities, at shops, offices, and warehouses. The same requirements must be observed as in section 2.18, "Parallel Operation of Single-phase Transformers for Additional Capacity," with one additional requirement—in the manner connections are made between individual single-phase units. ANSI standard connections are illustrated below in the following figures. There are other angular displacements that will work but are seldom used. Do not attempt to connect single-phase units together in any combination that does keep the exact angular displacement on both primary and secondary; a dangerous short circuit could be the result. Additive and subtractive polarities can be mixed (see the following figures). These banks also may be paralleled for additional capacity if the rules are followed for three-phase paralleling discussed below. When paralleling individual three-phase units or single-phase banks to operate three phase, angular displacements must be the same.

Figure 18 shows delta-delta connections. Figure 19 shows wye-wye connections, which are seldom used at Reclamation facilities, due to

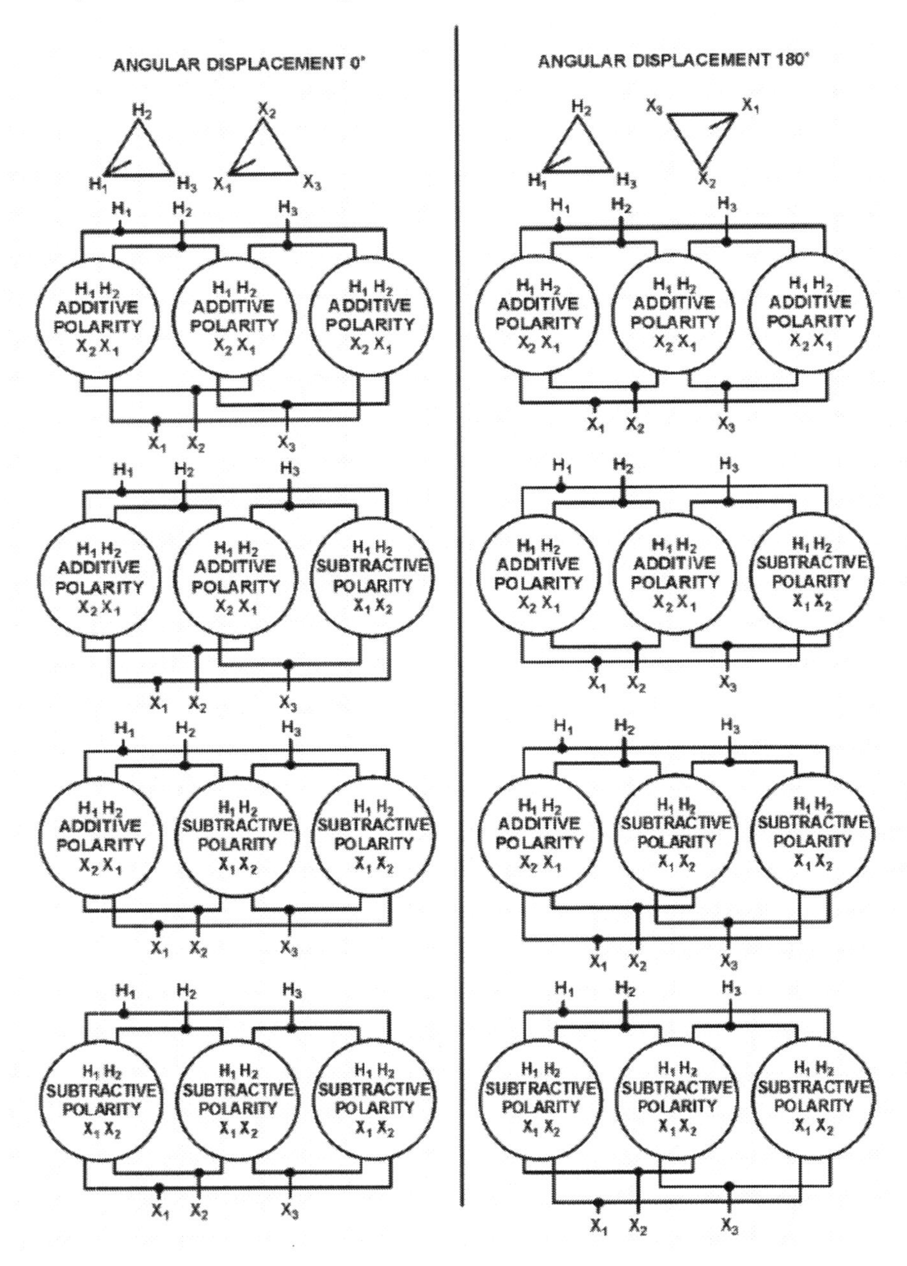

Figure 18 – Delta-Delta Connections, Single-Phase Transformers for Three-Phase Operation.

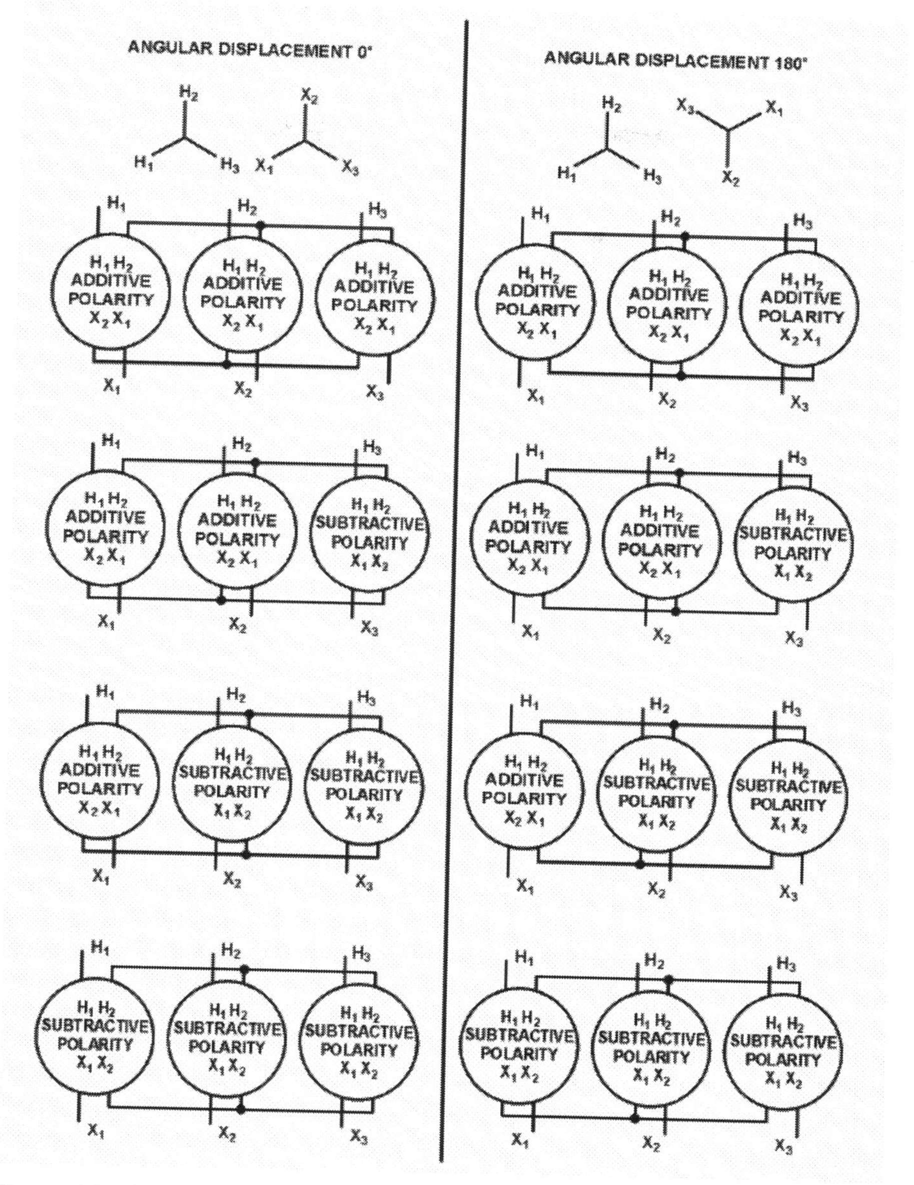

Figure 19 – Wye-Wye Connections, Using Single-Phase Transformers for Three-Phase Operation.

Note: These connections are seldom used because of 3rd harmonic problems.

inherent third harmonic problems. Methods of dealing with the third harmonic problem by grounding are listed below.

However, it is easier just to use another connection scheme (i.e., delta-delta, wye-delta, or delta-wye [figure 20]), to avoid this problem altogether. In addition, these schemes are much more familiar to Reclamation personnel.

2.22 Paralleling Three-Phase Transformers

Two or more three-phase transformers, or two or more banks made up of three single-phase units, can be connected in parallel for additional capacity. In addition to requirements listed above for single-phase transformers, phase angular displacements (phase rotation) between high and low voltages must be the same for both. The requirement for identical angular displacement must be met for paralleling any combination of three-phase units and/or any combination of banks made up of three single-phase units.

CAUTION:

This means that some possible connections will not work and will produce dangerous short circuits. See table 2 below.

For delta-delta and wye-wye connections, corresponding voltages on the high-voltage and low-voltage sides are in phase. This is known as zero phase (angular) displacement. Since the displacement is the same, these may be paralleled. For delta-wye and wye-delta connections, each low-voltage phase lags its corresponding high-voltage phase by 30 degrees. Since the lag is the same with both transformers, these may be paralleled. A delta-delta, wye-wye transformer, or bank (both with zero degrees displacement) cannot be paralleled with a delta-wye or a wye-delta that has 30 degrees of displacement. This will result in a dangerous short

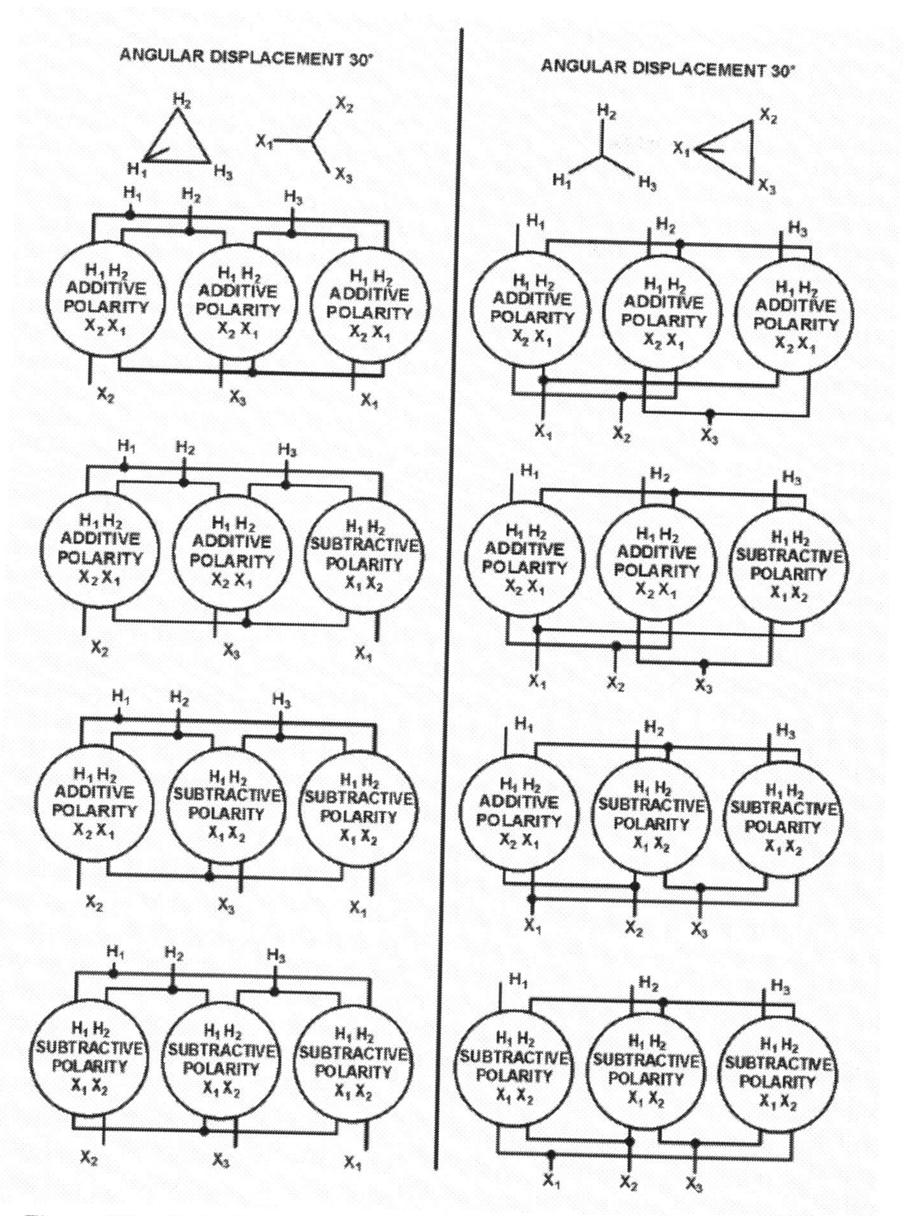

Figure 20 – Delta-Wye and Wye-Delta Connections Using Single-Phase Transformers for Three-Phase Operation.

Note: Connections on this page are the most common and should be used if possible.

circuit. Table 1 shows the combinations that will operate in parallel, and table 2 shows the combinations that will not operate in parallel.

Table 1 – Operative Parallel Connections of Three-Phase Transformers

OPERATIVE PARALLEL CONNECTIONS				
	LOW-VOLTAGE SIDES		HIGH-VOLTAGE SIDES	
	Trans. A	Trans. B	Trans. A	Trans. B
1	Delta	Delta	Delta	Delta
2	Y	Y	Y	Y
3	Delta	Y	Delta	Y
4	Y	Delta	Y	Delta
5	Delta	Delta	Y	Y
6	Delta	Y	Y	Delta
7	Y	Y	Delta	Delta
8	Y	Delta	Delta	Y

Table 2 – Inoperative Parallel Connections of Three-Phase Transformers

INOPERATIVE PARALLEL CONNECTIONS				
	LOW-VOLTAGE SIDE		HIGH-VOLTAGE SIDE	
	Trans. A	Trans. B	Trans. A	Trans. B
1	Delta	Delta	Delta	Y
2	Delta	Delta	Y	Delta
3	Y	Y	Delta	Y
4	Y	Y	Y	Delta

Wye-wye connected transformers are seldom, if ever, used to supply plant loads or as GSU units, due to the inherent third harmonic problems with this connection. Delta-delta, delta-wye, and wye-delta are used extensively at Reclamation facilities. Some rural electric associations use wye-wye connections that may be supplying

Reclamation structures in remote areas. There are three methods to negate the third harmonic problems found with wye-wye connections:

1. Primary and secondary neutrals can be connected together and grounded by one common grounding conductor.
2. Primary and secondary neutrals can be grounded individually using two grounding conductors.
3. The neutral of the primary can be connected back to the neutral of the sending transformer by using the transmission line neutral.

In making parallel connections of transformers, polarity markings must be followed. Regardless of whether transformers are additive or subtractive, connections of the terminals must be made according to the markings and according to the method of the connection (i.e., delta or wye).

CAUTION:

As mentioned above regarding paralleling single-phase units, when connecting additive polarity transformers to subtractive ones, connections will be in different locations from one transformer to the next.

2.23 Methods of Cooling

Increasing the cooling rate of a transformer increases its capacity. Cooling methods must not only maintain a sufficiently low average temperature but must prevent an excessive temperature rise in any portion of the transformer (i.e., it must prevent hot spots). For this reason, working parts of large transformers are usually submerged in high-grade insulating oil. This oil must be kept as free as possible from moisture and oxygen, dissolved combustible gases, and particulates.

Figure 21 – Cooling.

Ducts are arranged to provide free circulation of oil through the core and coils; warmer and lighter oil rises to the top of the tank, cooler and heavier oil settles to the bottom. Several methods have been developed for removing heat that is transmitted to the transformer oil from the core and windings (figure 21).

2.24 Oil-Filled – Self-Cooled Transformers

In small- and medium-sized transformers, cooling takes place by direct radiation from the tank to surrounding air. In oil-filled, self-cooled types, tank surfaces may be corrugated to provide a greater radiating surface. Oil in contact with the core and windings rises as it absorbs heat and flows outward and downward along tank walls, where it is cooled by radiating heat to the surrounding air. These transformers may also have external radiators attached to the tank to provide greater surface area for cooling.

2.25 Forced-Air and Forced-Oil-Cooled Transformers

Forced-air-cooled transformers have fan-cooled radiators through which the transformer oil circulates by gravity, as shown in figure 22a. Fans force air through radiators, cooling the oil.

Forced-air/oil/water-cooled transformers have a self-cooled (kVA or MVA) rating and one or more forced cooling ratings (higher kVA or MVA). Higher ratings are due to forced cooling in increasing amounts. As temperature increases, more fans or more oil pumps are turned on automatically.

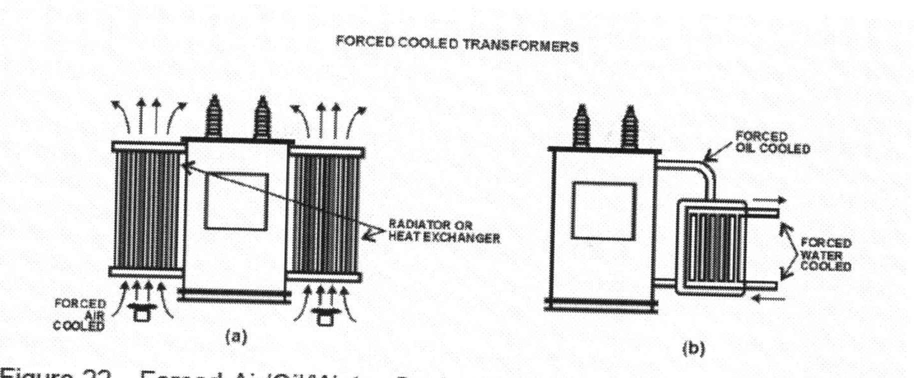

Figure 22 – Forced-Air/Oil/Water-Cooled Transformers.

The forced-cooling principle is based on a tradeoff between extra cooling and manufacturing costs. Transformers with forced-cooling have less weight and bulk than self-cooled transformers with the same ratings. In larger-sized transformers, it is more economical to add forced cooling, even though the electricity needed to operate fans and pumps increases the operating cost.

2.26 Transformer Oil

In addition to dissipating heat due to losses in a transformer, insulating oil provides a medium with high dielectric strength in which the coils and core are submerged. This allows the transformers to be more compact, which reduces costs.

Insulating oil in good condition will withstand far more voltage across connections inside the transformer tank than will air. An arc would jump across the same spacing of internal energized components at a much lower voltage if the tank had only air. In addition, oil conducts heat away from energized components much better than air.

Over time, oil degrades from normal operations, due to heat and contaminants. Oil cannot retain high dielectric strength when exposed to air or moisture. Dielectric strength declines with absorption of moisture and oxygen. These contaminants also deteriorate the paper insulation. For this reason, efforts are made to prevent insulating oil from contacting air, especially on larger power transformers. Using a tightly sealed transformer tank is impractical, due to pressure variations resulting from thermal expansion and contraction of insulating oil. Common systems of sealing oil-filled transformers are the conservator with a flexible diaphragm or bladder or a positive-pressure inert-gas (nitrogen) system. Reclamation GSU transformers are generally purchased with conservators, while smaller station service transformers have a pressurized nitrogen blanket on top of oil. Some station service transformers are dry-type, self-cooled or forced-air cooled.

2.27 Conservator System

A conservator is connected by piping to the main transformer tank that is completely filled with oil. The conservator also is filled with oil and contains an expandable bladder or diaphragm between the oil and air to prevent air from contacting the oil. Figure 23 is a schematic

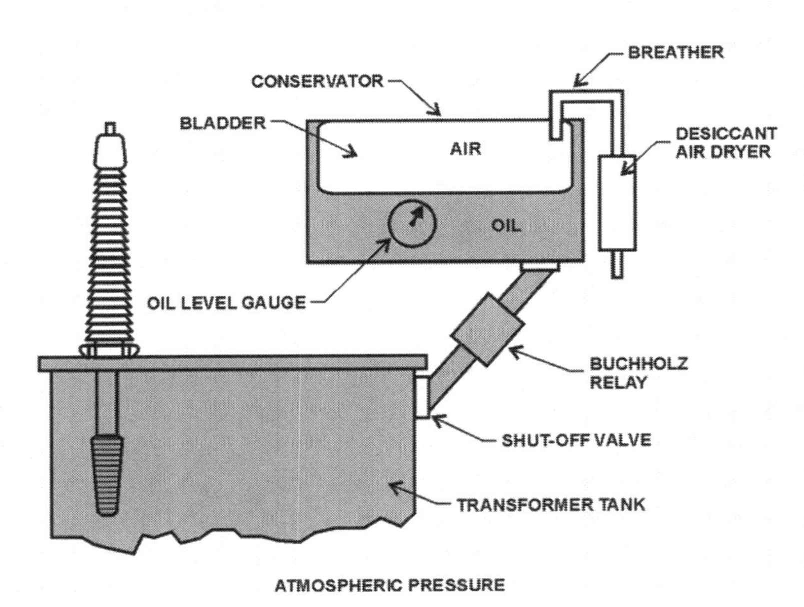

Figure 23 – Conservator with Bladder.

representation of a conservator system (figure 1 is an actual photo of a conservator). Air enters and exits the space above the bladder/ diaphragm as the oil level in the main tank goes up and down with temperature. Air typically enters and exits through a desiccant-type air dryer that must have the desiccant replaced periodically. The main parts of the system are the expansion tank, bladder or diaphragm, breather, vent valves, liquid-level gauge and alarm switch. Vent valves are used to vent air from the system when filling the unit with oil. A liquid-level gauge indicates the need for adding or removing transformer oil to maintain the proper oil level and permit flexing of the diaphragm. These are described in detail in section 4.4.

2.28 Oil-Filled, Inert-Gas System

A positive seal of the transformer oil may be provided by an inert-gas system. Here, the tank is slightly pressurized by an inert gas such as nitrogen.

The main tank gas space above the oil is provided with a pressure gauge (figure 24). Since the entire system is designed to exclude air, it must operate with a positive pressure in the gas space above the oil; otherwise, air will be admitted in the event of a leak. Smaller station service units do not have nitrogen tanks attached to automatically add gas, and it is common practice to add nitrogen yearly each fall as the tank starts to draw partial vacuum, due to cooler weather. The excess gas is expelled each summer as loads and temperatures increase.

Some systems are designed to add nitrogen automatically (figure 24) from pressurized tanks when the pressure drops below a set level. A positive pressure of approximately 0.5 to 5 pounds per square inch (psi) is maintained in the gas space above the oil to prevent ingress of air. This system includes a nitrogen gas cylinder; three-stage, pressure-reducing valve; high-and low-pressure gauges; high- and low-pressure alarm switch; an oil/condensate sump drain valve; an automatic pressure-relief valve; and necessary piping.

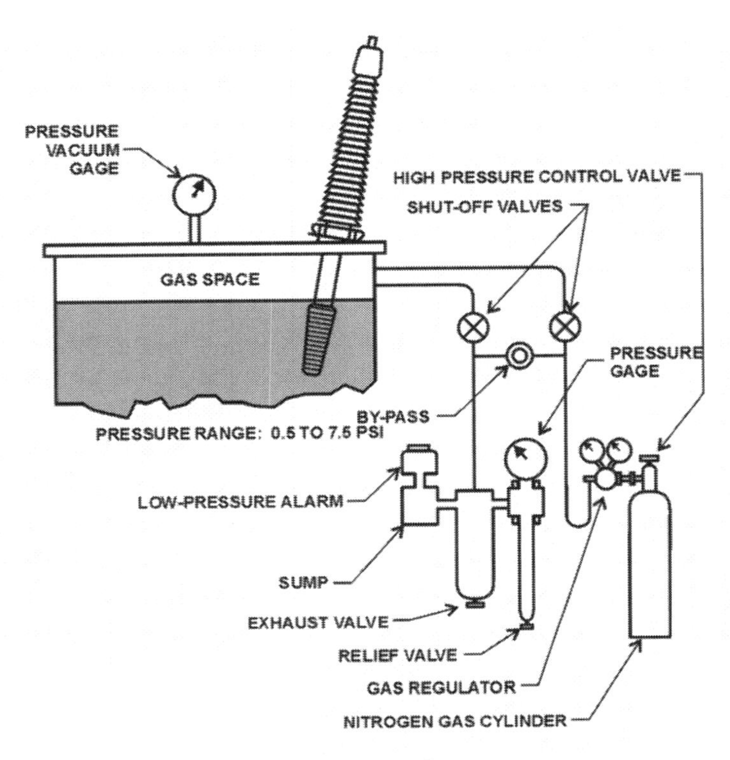

Figure 24 – Typical Transformer Nitrogen System.

The function of the three-stage, automatic pressure-reducing valves is to reduce the pressure of the nitrogen cylinder to supply the space above the oil at a maintained pressure of 0.5 to 5 psi.

The high-pressure gauge normally has a range of 0 to 4,000 psi and indicates nitrogen cylinder pressure. The low-pressure gauge normally has a range of about -5 to +10 psi and indicates nitrogen pressure above the transformer oil.

In some systems, the gauge is equipped with high- and low-pressure alarm switches to alarm when gas pressure reaches an abnormal value; the high-pressure gauge may be equipped with a pressure switch to sound an alarm when the supply cylinder pressure is running low.

A sump and drain valve provide a means for collecting and removing condensate and oil from the gas. A pressure-relief valve opens and closes to release the gas from the transformer and, thus, limit the pressure in the transformer to a safe maximum value. As temperature of a transformer rises, oil expands, and internal pressure increases, which may have to be relieved. When temperature drops, pressure drops, and nitrogen may have to be added, depending on the extent of the temperature change and pressure limits of the system. The pressurized gas system is discussed in detail in section 4.9.1.3.

2.29 Indoor Transformers

When oil-insulated transformers are located indoors, because of fire hazard, it is often necessary to isolate these transformers in a fireproof vault.

Today, dry-type transformers are used extensively for indoor installations. These transformers are cooled and insulated by air and are not encased in sealed tanks like liquid-filled units. Enclosures in which they are mounted must have sufficient space for entrance, for circulation of air, and for discharge of hot air. Dry-type transformers are enclosed in sheet metal cases with a cool air entrance at the bottom and a hot air discharge near the top. They may or may not have fans for increased air flow.

In addition to personnel hazards, indoor transformer fires are extremely expensive and detrimental to plants, requiring extensive cleanup, long outages, and lost generation. Larger indoor transformers, used for station service and generator excitation, should have differential relaying so that a fault can be interrupted quickly before a fire can ensue. Experience has shown that transformer protection by fuses alone is not adequate to prevent fires in the event of a short circuit.